YOUR KNOWLEDGE HAS VALUE

- We will publish your bachelor's and master's thesis, essays and papers

- Your own eBook and book - sold worldwide in all relevant shops

- Earn money with each sale

Upload your text at www.GRIN.com and publish for free

Bibliographic information published by the German National Library:

The German National Library lists this publication in the National Bibliography;
detailed bibliographic data are available on the Internet at http://dnb.dnb.de .

Imprint:

Copyright © 2016 GRIN Verlag, Open Publishing GmbH
Print and binding: Books on Demand GmbH, Norderstedt Germany
ISBN: 9783668478602

This book at GRIN:

http://www.grin.com/en/e-book/370601/isolation-characterization-and-optimization-
of-dye-degrading-bacteria

Prem Jose Vazhacharickal, John Joseph, Jiby John Mathew, Sajeshkumar N.K., Delmy Abraham

Isolation, characterization and optimization of dye degrading bacteria from natural source

An overview

GRIN Publishing

GRIN - Your knowledge has value

Since its foundation in 1998, GRIN has specialized in publishing academic texts by students, college teachers and other academics as e-book and printed book. The website www.grin.com is an ideal platform for presenting term papers, final papers, scientific essays, dissertations and specialist books.

Visit us on the internet:

http://www.grin.com/

http://www.facebook.com/grincom

http://www.twitter.com/grin_com

Isolation, characterization and optimization of dye degrading bacteria from natural source: an overview

Prem Jose Vazhacharickal, John Joseph, Jiby John Mathew,

Sajeshkumar N.K, and Delmy Abraham

ACKNOWLEDGEMENTS

Firstly we thank **God Almighty** whose blessing were always with us and helped us to complete this project work successfully.

We wish to thank our beloved Manager **Rev. Fr. Dr. George Njarakunnel,** Respected Principal **Dr. Joseph V.J,** Vice Principal **Fr. Joseph Allencheril,** Bursar **Shaji Augustine** and the Management for providing all the necessary facilities in carrying out the study. We express our sincere thanks to **Mr. Binoy A Mulanthra** (lab in charge, Department of Biotechnology) for the support. This research work will not be possible with the co-operation of many farmers.

Lastly, we extend our indebt thanks to patents, friends, and well wishers for their love and support.

Prem Jose Vazhacharickal*, John Joseph, Jiby John Mathew, Sajeshkumar N.K and Delmy Abraham

*Address for correspondence
Assistant Professor
Department of Biotechnology
Mar Augusthinose College
Ramapuram-686576
Kerala, India

Table of contents

Table of figures

Table of tables

List of abbreviations

%	: Percentage
°C	: Degree celsius
µl	: Microlitre
AOP	: Advanced oxidation process
CFU	: Colony forming unit
CO_2	: Carbon dioxde
COD	: Chemical oxygen demand
g/L	: Gram per litre
H_2O	: Water
H_2O_2	: Hydrogen peroxide
HCl	: Hydrochloric acid
Mg/L	: Milligram per litre
mL	: Millilitre
NaCl	: Sodium chloride
NaOH	: Sodium hydroxide
nm	: Nanometre
O_3	: Ozone
OD	: Optical density
RO	: Reverse osmosis
SE	: Standard error
Sp	: Species
TSI	: Triple Sugar Iron
UV	: Ultraviolet
λmax	: Lamda maximum
ZVI	: Zero valent iron

Isolation, characterization and optimization of dye degrading bacteria from natural source: an overview

Prem Jose Vazhacharickal[1], John Joseph[2], Jiby John Mathew[1], Sajeshkumar N.K[1] , Delmy Abraham[2], Bilu Kurian[2], and Sreelakshmi V.P[2]

[1]Department of Biotechnology, Mar Augusthinose College, Ramapuram, Kerala, India-686576

[2]Department of Bioscience, Indira Gandhi College of Arts and Science, Nellikuzhi, Kerala, India-686691

Abstract

In this study an attempt was made to evaluate the colour degradation capabilities by collecting the contaminated soil sample from Kalady area and serial dilution was done upto 10^{-6}. From the dilution 10^{-5} was taken and spread plated on Nutrient agar. From the above plate, isolated colonies was obtained which was found to be Bacillus sp and Pseudomonas sp respectively by morphological, microscopical and biochemical method. The isolated colonies was taken for degradation studies with 1% dye and 1% inoculum in Nutrient broth and OD values and colour change was noted. It was found to be Bacillus sp has more degrading capacity in yellow colour than *Pseudmonas sp.* The optimization studies were done with Bacillus sp having different concentration of colour (2, 4, 6) with varying pH (4,6,8) and temperature (37°C, 40°C and room temperature). The result was found to be having the concentration of colour with 4% having pH 4 and temperature 37°C.

Keywords: Azo dyes; Dye degradation; Biochemical identification.

1. Introduction

India's dye industry produces every type of dyes and pigments. Production of dye stuff and pigments in India is close to 80,000 tones. India is the second largest exporter of dyestuffs and intermediates (developing countries) after China. The textile industry accounts for the largest consumption of dyestuffs, at nearly 80% (Carliell et al., 1995). Industrialization is vital to nation's economy because it serves as a vehicle for development. However, there are associated problems resulting from the introduction of industrial waste products into the environment. Many of these products are problematic because of persistence (low biodegradability) and/or toxicity.

Environmental pollution has been identified as a major problem in the modern world. The increasing demand for drinkable water, and its dwindling supply, has made the treatment and reuse of industrial effluents an attractive option. One of the most important environmental pollution problems is the colour in water courses; although some of these colours are normally present and of "natural" origins (colour originates from the activity of microorganisms in ponds), a considerable proportion, especially conurbation, originates from industrial effluents are associated with the production and use of dyes.

Azo dyes, the largest chemical class of dyes with the greatest variety of colours, have been used extensively for textile, dyeing, and paper painting (Carliell et al., 1995). These dyes cannot be easily degraded, and some are toxic to higher animals. Over 7×105 metric tonnes of synthetic dyes are produced worldwide every year for dyeing and printing, and out of this, about 5% - 10% are discharged with wastewater. The amount of dye lost depends on the class of dye applied: it varies from 2% loss with the use of basic dyes to about 50% loss in certain reactive sulfonated dyes (Dafale et al., 2008). The presence of dyes in aqueous ecosystem diminishes photosynthesis by impeding light penetration into deeper layers thereby deteriorating water quality and lowering the gas solubility. Furthermore, the dyes and their degraded by-products may be toxic to flora and fauna (Talarposhti, 2001). Azo dyes consist of diazotized amine coupled with an amine or phenol, and contain one or more azo linkages. At least 300 different varieties of azo dyes are extensively used in the textile, paper, food, cosmetics and pharmaceutical industries. The effect of pH, temperature, type and concentration of respiration substrates, and oxygen tension on the rate of biological reduction of a variety of azo dyes have previously been investigated (Wuhrmann et al., 1980; Deng et al., 2008). Several combinations of treatment methods have been developed so far in order to effectively process cotton-textile wastewater, with decolorization among the main goals for these processes. Chemical coagulation/flocculation techniques, usually combined with activated sludge treatment, have been among the most common processing methods mainly due to the ease of their application. However, the above methods require high amounts of raw materials [coagulants] and also yield large amount of waste solids, leading to the elevation of the total treatment cost. Advanced oxidation processes (Ozonation, UV/H_2O_2) are based on the generation of hydroxyl radicals, which are mainly highly reactive oxidants. They are environmental friendly techniques, since no solid wastes are produced. However, they are not cost effective due to the high consumption of both energy and raw material. Active carbon adsorption and nano-filtration techniques are able to remove dyes from wastewater. Although, the main disadvantage of these methods is the production of a secondary waste stream (or waste solid) that requires further treatment or disposal.

Microorganisms can play a very significant role in decomposition and ultimate mineralization of these dyes (Razo-Flores et al., 1996). Environmental biotechnology is based on ability of microorganism (both bacterial and fungal) to decompose larger chemical compounds, which are xenobiotics. Many researchers have studied in detail and have isolated several microbial strains having potential to decolorize a large number of dyes belonging to different classes have been isolated (Levine, 1991). Biodegradation of reactive azo dyes present in textile wastewater is a complicated procedure due to versatility in structure of dyes. The factors such as temperature, dissolved oxygen, redox mediators, type of microorganisms and amount of nutrients, type and chemical structure of dye are under study. Many research reports are available, which explain successful decolorization of dyes by using purified microbial cultures. But these findings do not find much application in practical treatment system due to complexity and heterogeneity of chemical compounds present in textile wastewater. Over the Past decades, Biological decolorization has been investigated as a method to transform, degrade or mineralize azo dyes. Moreover, such decolorization and degradation is an environmentally friendly and cost competitive alternative to chemical decomposition process. Unfortunately, most azo dyes are recalcitrant to aerobic degradation by bacterial cells. However, there are few known microorganisms that have the ability to reductively cleave azo bonds under aerobic conditions.

It was reported that cell extracts or membrane – permeabilized cells reduce azo dyes more efficiently than entire cells, especially in the case of sulphonatedazo dyes (Mezohegyi, et al., 2012). The flavin reductases are located in the cytoplasm of the cells (Alaton and Balcioglu, 2001), wh ch suggests that anaerobic reduction of azo dyes is an intracellular process and the permeation of azo dyes across the cell membrane is a rate limiting factor. On the contrary, using different bacterial cultures, other authors found similar removal rates for polymeric azo dyes and other analogue compounds with much simpler structures (Olukanni et al., 2006) an extracellular mechanism for azo dye reduction being proposed.

1.1 Objectives

The objectives of this study to isolate and characterize dye degrading bacteria and its evaluation.

2. Review of literature

2.1 Reactive dyes

A dye is described as a coloured substance with affinity to substrate applied. Dyes are soluble at some stage of the application process, whereas pigments in general retain basically their particulate or crystal ine form during application. These are used to impart colour to materials of which it becomes an integral part. Aromatic ring structure coupled with a side chain is usually required for resonance and in turn imparts colour. Based on the origin and complex molecular structure, dyes can be classified into three categories: (1) An onic: acid, direct and reactive dyes; (2) Cationic: basic dyes; and (3) Non-ionic: disperse dyes (Gong et al., 1993; Mishra and Tripathy, 1993; Fu & Viraraghavan, 2001; Greluk and Hubicki, 2010). It has been estimated that over 10,000 different textile dyes and pigments were in common use (Easton, 1995; Robinson et al., 2001). Also it is reported that there are over 100,000 commercial dyes are available with a production of over 7×105 metric tons per year (Zollinger, 1987; Fu and Viraraghavan, 2001) Among the various classes of dyes, reactive dyes are one of the prominent and most widely used types of azo dyes and are too difficult to eliminate. They are extensively used in different industries, including rubber, textiles, cosmetics, paper, leather, pharmaceutical and food (Aksu & Donmez, 2005; Vijayaraghavan & Yun, 2008; Wang et al., 2009). Because these dyes have favourable characteristics, such as wide colour spectrum , bright colour and colour shades, high wet fastness profiles, ease of application, brilliant colours and minimum energy consumption (Lee & Pavlostathis, 2004; Aksu, 2005; Vijayaraghavan and Yun, 2008).

The most common group reactive dyes are azo, anthraquinone, phthalocyanine (Axelsson et al., 2006) and reactive group dyes (Lin and Peng, 1994; Sanghi et al., 2006; Daneshvar et al., 2007). Most of these dyes are toxic and carcinogenic (Acuner and Dilek, 2004). Disposal of these dyes into the environment causes serious damage, like they may significantly affect the photosynthetic activity of hydrophytes by reducing light penetration (Aksu et al., 2007) and also they may be toxic to some aquatic organisms due to their breakdown products (Hao et al., 2000; He et al., 2007). Once they are released, they not only produce toxic amines by reductive cleavage of azo linkages which causes severe effects on human beings through damaging the vital organs such as brain, liver, kidneys, central nervous and reproductive systems (Aksu, 2005; Iscen et al., 2007) and light penetration (Brown and De Vito, 1993; O'mahony et al., 2002; Yesilada et al., 2003; Forgacs et al., 2004; Kalyani et al., 2008) in aquatic environment. Therefore, their removal causes a big environmental concern in industrialized countries and is subjected to many scientific researches. It is estimated that 10–20% of reactive dyes remain in wastewater during the production and nearly 50% of reactive dyes are lost through hydrolysis during the dyeing process and their removal from effluent is difficult by conventional physical/chemical as well as biological treatment (Manu and Chaudhari, 2002; Li et al., 2009; Greluk and Hubicki, 2010). Therefore, a large quantity of the dyes appears in wastewater (Heinfling et al., 1997). These dyestuffs are designed to resist biodegradation.

Synthetic reactive dyes are considered as recalcitrant xenobiotic compounds, due to the presence of an N=N bond and groups such as aromatic rings that are not easily degraded. The discharge of these coloured compounds into the environment causes considerable non-aesthetic pollution and serious health risks (Martínez Huitle and Brillas, 2009).

2.2 Removal methods

Many processes were employed to remove dye molecules from industry effluents and the treatment methods can be divided into the following categories:

2.3 Physical methods

Physical methods such as Adsorption (Chatterjee et al., 2010a; Chatterjee et al., 2010b), Ion exchange (Labanda et al., 2009) and Membrane filtration (Ahmad and Puasa, 2007) were employed in the removal of dyes. The main disadvantages of these physical methods were they simply transfer the dye molecules to another phase rather than destroying them and they were effective only when the effluent volume is small (Robinson et al., 2001).

2.4 By adsorption

Adsorption is the transfer of solute dye molecule at the interface between two immiscible phases in contact with one another. The removal of colour from dye industrial effluents by the adsorption process using granular activated carbon has emerged as a practical and economical approach.

2.5 By ion exchange

Removal of anions and cations from dye industry effluent can be carried out by Ion exchange method by passing the waste water through the beds of ion exchange resins where some undesirable cations or anions of waste water get exchanged for sodium or hydrogen ions of the resin. Greluk and Hubicki (2010) recommended the adsorption/ion exchange as an alternative method for the removal of reactive dyes. Application of commercial anion exchange resins to water contaminated with a broad range of reactive dyes were studied by Karcher et al. (2001) and reported that anion exchangers possess excellent adsorption capacity (200–1200 µmol/g) as well as efficient regeneration property for their removal and recovery. The applicability of ion exchange resin containing acrylic matrix for removing other classes of dyes were well documented by Bayramoglu et al. (2009), Dulman et al. (2009), Wawrzkiewicz and Hubicki (2009) and Barsanescu et al. (2009). Thus Acrylic anion exchangers is more advantage than styrenics by exhibiting high efficiency of anion exchange capacities and polluting less.

2.6 By membrane filtration

Reverse osmosis (RO) and electro dialysis are the important examples of membrane filtration technology. Electrolyte is important in dyeing process for exhaustion of dye. The concentration of neutral electrolyte like sodium chloride (NaCl) in the dyeing bath is in the range of 25/30 g/L for deep tone, 41.5 g/L for light tone and extended to 50 g/L in some exceptional cases. The exhaustion stage in reactive dyeing on cotton also requires sufficient quantity of salt. The contribution of reverse osmosis in removing this high salt concentration is of great. This RO reject can be reused again in the process. For reactive dyeing on cotton, the presence of electrolytes in the waste water causes an increase in the hydrolysed dye affinity making it difficult to extract. The total dissolved solids from waste water were removed by reverse osmosis. Though it is suitable for removing ions and larger species from dye bath effluents with high efficiency, it possesses some disadvantages like clogging of the membrane by dyes after long usage and high capital cost. In electro dialysis, the dissolved salts (ionic in nature) can also be removed by impressing an electrical potential across the water, resulting in the migration of cations and anions to respective electrodes via anionic and cationic permeable membranes.

To avoid membrane fouling it is essential that turbidity, suspended solids, colloids and trace organics are to be removed prior to electro dialysis.

2.7 Chemical methods

Chemical methods such as chemical oxidation (Osugi et al., 2009), electrochemical degradation (Yi et al., 2008), and ozonation (Moussavi & Mahmoudi, 2009) were employed in dye removal effectively. The treatment of synthetic dye house effluent byozonation and hydrogen peroxide in combination with Ultraviolet light was vast in literature. A variety of oxidizing agents were used to decolorize wastes by oxidation techniques effectively. Among that sodium hypochlorite decolorizes dye bath efficiently. Even though it is a low cost technique, it forms absorbable toxic organic halides. Ozone on decomposition generates oxygen and free radicals. The later combines with coloring agents of effluent, resulting in the destruction of colors. The main disadvantage of this technique is that it requires an effective sludge producing pre-treatment. Also, these chemical methods with high cost were rarely used in the actual treatment process and the disposal of sludge containing chemicals at the end of treatment requires further use of chemicals (Crini, 2006; Forgacs et al., 2004).

2.8 Advanced oxidation process (AOP)

Philippe et al. (1998); Slokar & Le Marechal (1998) were reported that the conventional water treatment technologies such as solvent extraction, activated carbon adsorption and chemical treatment process such as oxidation by ozone (O_3) often produce hazardous by-products and generate large amount of solid wastes, which require costly disposal or regeneration method. Due to these reasons, Mahadwad et al. (2011) considerable attention had been focused on complete oxidation of organic compounds to harmless products such as carbon dioxide (CO_2) and water (H_2O) by the AOP. El-Dein et al. (2003) supported the AOP and reported that it provides a promising alternative method to treat the textile wastewater. The UV-driven AOPs use UV light with an oxidizer such as H_2O_2 and/or ozone to generate hydroxyl radicals (OH^-) that attack organic compounds non selectively with a high reaction rate. Based on the studies from Shu et al. (1994) and Galindo & Kalt (1998) it was observed that the decolorization of textile dyes using H_2O_2/UV had shown it to decolorize dilute aqueous solutions (20 mg/L) of azo dyes.

2.9 Electrochemical Method

The requirement of chemicals and the temperature to carry the electro chemical reaction is less than those of other equivalent non-electrochemical treatment. It can also prevent the production of unwanted side products. But, if suspended or colloidal solids were high in concentration in the waste water, they slow down the electrochemical reaction. Therefore, those materials need to be sufficiently removed before electrochemical oxidation. Ceron et al. (2004) reported that many of the commercially used dyes are resistant to biological and physico-chemical methods (Delee et al., 1998; (Vandevivere et al., 1998; Anbia et al., 2010). Also Ceron et al. (2004) suggested that coagulation (Vancevivere et al., 1998), coagulation – electro oxidation (Xiong et al., 2001), adsorption (Morais et al., 1999), electrolysis (Davila-Jimenez et al., 2000photolysis (Ince, 1999) and ozonation are promising in terms of performance. But in terms of economic aspect, these methods have become most challenging problem. Consequently, Gutierrez et al. (2001) discussed the interest in electrochemical methods to decolourise and degrade dye molecules. The electric current induces redox reactions resulting in the transformation and destruction of the organic compounds and almost complete oxidation to CO_2 and H_2O. At the same time Powell et al. (1994) reported that the small scale oxidizing methods using Fenton's reagent, which has lower costs in comparison with ozone process in dye liquor treatment. The oxidizing effect of the corona discharge is also known and it has been reported by Goheen et al. (1994) as an effective method to bleach organic dyes using a stainless steel electrode. Steel electrodes were used commonly in electrochemical technology to remove color by generating ferrous hydroxide and ferric oxyhydroxide. Yang et al. (2000) reported a new result for the color removal of dye from wastewater by applying electro generated hypochlorite ions and (Ru+ Pt) Ox binary electrodes. Even if the removal of dyes from wastewater in an economic way by using electrochemical method, the low-cost electrode production remains a major concern. Zero Valent Iron (ZVI) Chatterjee et al. (2009) discussed the development of new treatment strategies to degrade the dye molecules using ZVI particles. These are inexpensive, environmental friendly strong reducing agents and Sun et al. (2006) reported that ZVI can donate two electrons to many environmental contaminants as $FeO \rightarrow Fe^2 +$ 2e- Chatterjee et al. (2009) also reported that due to its effective electron donating capacity, ZVI particles had been studied for the treatment of wastewater contaminated with chlorinated compounds, nitro aromatic compounds, nitrates, heavy metals, organochlorine pesticides, and dyes. Fan et al. (2009) reported that the reaction between FeO and H_2O or H^+ can generate H atoms, which induce cleavage of the azo bond (–N N–), thus damaging the chromophore group and conjugated system of ulation and bithe azo dyes.Particles for the decolorization process include a low iron

Concentration remaining in the sludge, no requirement for further treatment of effluents and easy recycling of the spent iron powder by magnetism. Low-cost, is easy-to obtain, and has good effectiveness and ability of degrading Saxe about the other advantages of using ZVI particles convert azo dye into some products that were more susceptible to contaminants.

2.10 Biological methods
Biological degradation process. Chang et al. (2006) discussed Bioaccumulation and bio-sorption are the two main technologies in biological process for of dye bearing industrial effluents. They possess good potential to), replace conventional methods

for the treatment of dyes industry effluents (Volesky and Holan, 1995; Malik, 2004). Biological process can be carried out in situ at the contaminated site, these are usually environmentally benign i.e., no secondary pollution and they were cost effective. These are the principle advantages of biological technologies for the treatment of dye industry effluents. Hence in recent years, research attention has been focused greatly on biological methods for the treatment of effluents (Vara Prasad and de Oliveira Freitas, 2003; Vijayaraghavan and Yun, 2008). The disadvantage of this degradation process is that it suffers from low degradation efficiency or even no degradation for some dyes (Stolz, 2001; Pearce et al., 2003) and practical difficulty in continuous process (Vijayaraghavan & Yun, 2008) clearly demonstrated the difference between bio-accumo sorption in their review. Bioaccumulation is defined as the phenomenon of uptake of toxicants by living cells; whereas, bio-sorption can be defined as the passive uptake of toxicants by dead or inactive biological materials. The important advantage of bio-sorption than bioaccumulation process is the use of living organisms is not advisable for the continuous treatment of highly toxic effluents. This problem can be overcome by the use of dead biomass, which is flexible to environmental conditions and toxicant concentrations. Erdal and Taski (2010) discussed about various treatment methods exist for the removal of colour from industrial effluents, including physico-chemical and

biological processes. Although a number of chemical, physical processes namely flocculation, chemical coagulation, precipitation, ozonation and adsorption were employed for the treatment of dye bearing wastewaters. Aravindhan et al. (2007) and Sarioglu et al. (2007) reported that they possess some inherent limitations such as high cost, formation of hazardous by-products and intensive energy requirements. In addition to that Erdal and Taski (2010), Banat et al. (1996), Slokar and Marechal, (1998) reported that the physico-chemical processes were usually inefficient, costly and not adaptable to a wide range of dye wastewater. Erdal and Taski (2010), Fu & Viraraghavan (2001); Wang et al. (2009) and Aksu (2005) had suggested the increasing interest of biological processes, such as biodegradation, bioaccumulation and bio-sorption due to their cost effectiveness, ability to produce less sludge and environmental benignity. Fungi and algae had been played important role in dye decolorization. Wang et al. (2009) reported that adsorption rather than degradation plays a major role during the decolorization process by fungi and algae.

2.11 Anaerobic treatment

Karatas et al. (2010) reported the unsuitability of physicochemical decolorization methods regarding cost effectiveness, usage areas, interfere with other wastewater components, or cause wastes that require retreatment and also they suggested that the biologic treatment method especially anaerobic treatment in the case of azo dyes is an alternative to the physicochemical method which was relatively inexpensive and may be preferred for decolorization based on the investigation by Van der Zee et al. (2001); Carliell et al. (1995) supported the same i.e., in most cases, the dyes were easily reduced under anaerobic condition. The main disadvantage of azo dye reduction under anaerobic conditions were the production of aromatic amines, which usually do not degrade under these conditions (Mendez-Paz et al.,2005; Razo-Flores et al., 1996) and tend to accumulate a toxic levels (Carliell et al. 1995; Gottlieb et al.,2003). Such amines, however, were reported to be readily bio-transformed under aerobic conditions (Tan et al., 2000). The colour and chemical oxidation demand (COD) removal efficiencies were investigated by Sponza & Isýk (2002) using anaerobic - aerobic sequential processing for treatment of 100 mg/L of di-azo dye with glucose as the carbon source and reported the colour removal efficiency as 96%. Supaka et al. (2004) obtained 78.2% colour removal and 90% COD removal in a sequential anaerobic – aerobic system that was used to treat Remozal Black B dye. Sponza & Isýk (2002) reported 92.3 and 95.3% colour and COD removal efficiencies, respectively, when using an up flow anaerobic sludge blanket -aerobic stirred tank reactor sequential system to treat Congo Red dye. Kapdan and Oztekin, (2006) investigated Remozal Rot dye and reported over 90% colour removed and more than 85% COD removal efficiency in an anaerobic/aerobic SBR system. Khehra et al. (2006) reported 98% colour removal and 95% COD removal efficiency in an anoxicaerobic sequential bioreactor system used to treat Acid Red 88 azo dye. Zaoyan et al. (1992) obtained65% color and 74% COD removal efficiencies in textile wastewater contaminated with azo dyes using an anaerobic-aerobic rotating bio-disc system.

2.12 Enzymatic treatments

Roriz et al. (2009) mentioned that physical and chemical methods have high costs, low efficiency and cannot be used with a great variety of dyes and they suggested that the use of enzyme based methods are good alternative. Compared to the conventional methods, application to recalcitrant materials, operation at high and low contaminant concentrations over a wide pH, temperature and salinity range, biomass acclimatization was irrelevant and straight forward process control were the potential advantages of the enzymatic treatments reported by Duran & Esposito (2000) and Roriz et al. (2009).

3. Hypothesis

The current research work is based on the following hypothesis

1) Dye contaminated soil samples posse's diverse bacteria capable of utilizing various dyes.

2) These bacterial isolates differ in their dye degrading capabilities.

4. Materials and Methods

4.1 Study area

Kerala state covers an area of 38,863 km^2 with a population density of 859 per km^2 and spread across 14 districts. The climate is characterized by tropical wet and dry with average annual rainfall amounts to 2,817 ± 406 mm and mean annual temperature is 26.8°C (averages from 1871-2005; Krishnakumar et al., 2009). Maximum rainfall occurs from June to September mainly due to South West Monsoon and temperatures are highest in May and November.

4.2 Soil sample collection

Soil samples were collected from a region which was treated textile dye dump up area for several years. The soil samples were collected up 10 cm depth using standard soil collection protocols.

4.3 Quantitative analysis of microorganisms present in the collected soil

About 10 g of soil sample was diluted with 100 ml of distilled water. The soil sample was serially diluted to a dilution rate of 10^{-6} dilution. 25 ml nutrient agar plate was prepared and 0.1ml sample was spread plated (10^{-5} dilution). The plates were kept for 24 hrs incubation at 37°C in the bacteriological incubator. Colonies were counted on reaching the incubation time by the following equation:-

Total number of bacterial colony forming units in the sample

(CFU/ml) =   No. of colonies/ Volume of sample added* Dilution factor.

4.4 Isolation of potential dye decolourizers

1 ml of the sample was serially diluted to 10^{-5} from it. 0.1 ml from the sample serially diluted to 10^{-4} was spread plated and kept for 24 hr incubation at 37°C in the bacteriological incubator. Single colonies with distinct colony morphology were observed and each single colony was streaked onto nutrient agar plates to obtain pure cultures of isolates.

4.5 Pure culture preparation

Bacterial pure culture was prepared by streak plate method. One loop full of enrichment culture from the flasks was streaked on nutrient agar plates. The growth of the bacterial colonies was measured after 24-48 hr incubation at 28°C. Morphologically dissimilar colonies were randomly selected and sub cultured onto nutrient medium and maintained at 4°C for bacterial characterization.

4.6 Morphological and biochemical tests

Identification of the isolates were performed according to their morphological, cultural and biochemical characteristics by following Bergey's Manual of Systematic Bacteriology. All the isolates were subjected to Gram staining and specific biochemical tests.

4.6.1 Colony morphology

This was done to determine the morphology of selected strains on the basis of shape, size and colour.

4.6.2 Gram staining

A clean grease free slide was taken and a smear of the bacterial culture was made on it with a sterile loop. The smear was air-dried and then heat fixed. Then it was subjected to the following staining reagents:

(i) Flooded with Crystal violet for 1 min. followed by washing with running distilled water.

(ii) Again, flooded with Gram's Iodine for 1 min. followed by washing with running distilled water.

(iii) Then the slide was flooded with Gram's Decolourizer for 30 sec.

(iv) After that the slide was counter stained with Safranin for 30 sec, followed by washing with running distilled water.

(v) The slide was air dried and cell morphology was checked under microscope.

4.6.3 Motility test

This test is done to check the motility of the microorganism. The isolates were taken in a cavity slide from a 24 hr inoculated broth and observe under the microscope.

4.6.4 Sugar fermentation test

Various carbohydrates were used and added to peptone water to test the fermentation reaction. Gas production was detected by placing an inverted Durham's tube in the medium. The carbohydrates used include sucrose, lactose, and glucose. The result was read by the change in the colour from blue to yellow.

4.6.5 Oxidative fermentation test

The test is used to detect the oxidative fermentation characteristics of the test organisms. Certain bacteria use carbohydrates oxidatively and produce acids only in open tubes and contain other produced acids both in open and closed tubes fermentatively. Positive organisms showed colour change from blue to yellow.

4.6.6 Nitrate reduction test

It is used to study the ability of bacteria to reduce nitrate to nitrite. The organism that posses the enzyme nitrate reductase reduce nitrate to nitrite. Positive organism showed a pink colour when alpha naphthalamine and sulphanilic acid was added to the medium.

4.6.7 Triple sugar iron agar test

The test helps to determine whether the organisms utilize glucose, lactose or sucrose fermentatively and produce hydrogen sulphide. Yellow colour indicates acid production due to fermentation of lactose or sucrose or both. Pink colour indicates that the slant is alkaline, i.e. no fermentation of lactose or sucrose or both. Alkaline slant or acid butt indicates fermentation of glucose.

4.6.8 Coagulase test

Coagulase test determine *Staphylococcus aureus* from other members in the genus. S.aureusproduce an enzyme coagulase which activates prothrombin to clot rabbit plasma. The clot formation in the tube helps to identify the organism as *Staphylococcus aureus*.

4.6.9 Oxidase test

Oxidase test is used to detect the oxidase producing organisms. The cytochrome oxidase enzyme is able to oxidize the substrate tetra methyl paraphenylenediamenedihydrochloride forming a coloured product. Dark purple end product will be visible when a small amount of culture rubbed on the substrate impregnated on the paper.

4.6.10 Catalase test

It helps to demonstrate whether the bacteria are capable of producing catalase enzyme. Effervescence formation indicates positive result.

4.6.11 Mannitol motility test

The test is used to detect the ability of the test organism to ferment mannnitol and to check the motility of the organism. Positive organisms showed a change in the colour of the medium from red to yellow after incubation. Motile organisms showed diffused growth in the medium.

4.6.12 Indole production test

This test is conducted to distinguish the bacteria based on the ability to produce indole from tryptophan. Indole broth contains tryptophan rich with peptone and sodium chloride (NaCl). Formation of red ring within seconds on addition of Kovac's reagent shows a positive result.

4.6.13 Methyl red test

The test is used to study the ability of bacteria to produce sufficient acid during fermentation of glucose. Positive result shows a red colour on addition of methyl red indicator.

4.6.14 Voges-Proskauer test

VP test is conducted by to study the production of acetone during fermentative degradation of glucose. Positive organisms showed formation of pink colour on addition of VP agent.

4.6.15 Citrate utilization test

This test is used to study the ability of bacteria to utilize citrate as the sole source of carbon. Development of blue colour after 24 – 48 hours in the tube inoculated with positive organisms shows positive result.

4.6.16 Urease test

This test was used to detect the ability of bacteria to produce enzyme urease which hydrolyses urea and releases ammonia and carbon dioxide. Ammonia reacts in solution to form ammonium carbonate, which is alkaline and lead to the increase in pH. Phenol red changed its colour from yellow to red in alkaline pH thus indicating the presence of urease activity.

4.7 Statistical analysis

The survey results were analyzed and descriptive statistics were done using SPSS 12.0 (SPSS Inc., an IBM Company, Chicago, USA) and graphs were generated using Sigma Plot 7 (Systat Software Inc., Chicago, USA).

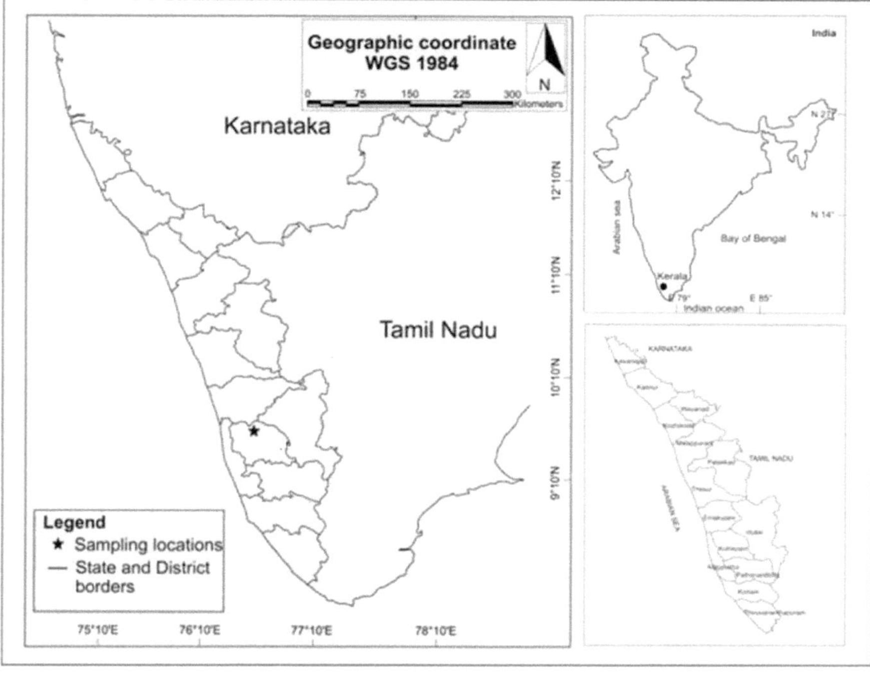

Figure 1. Map of Kerala showing the soil sample collection point. Authors own work.

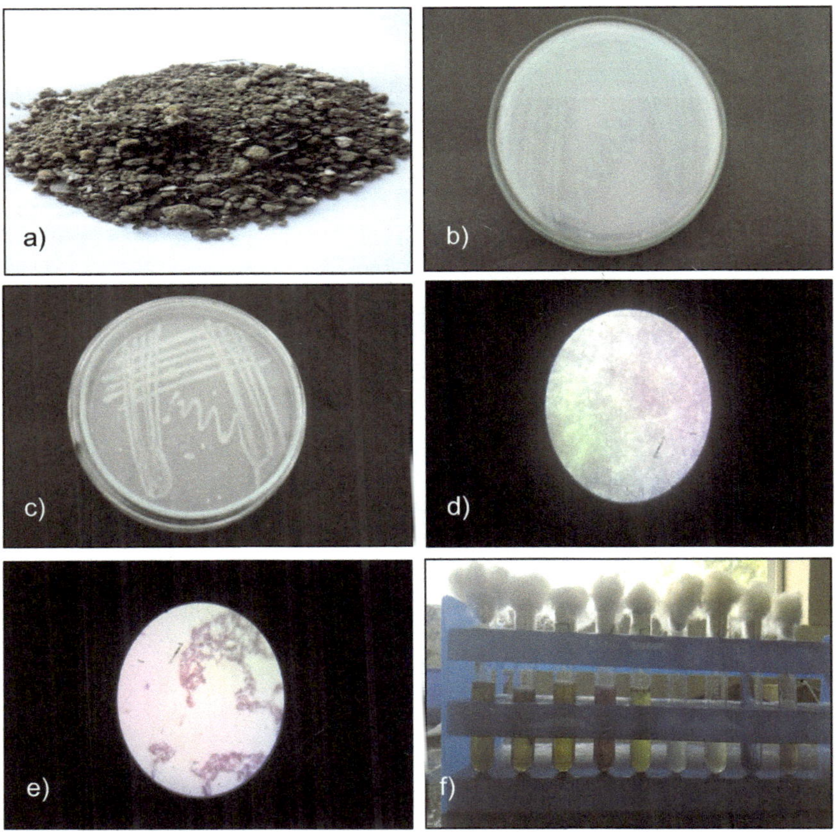

Figure 2. Details of a) soil sample collected, b) diluted soil samples containing bacterial isolate 1, c) diluted soil samples containing bacterial isolate 2, d) and e) gram staining of bacterial isolate 1 and 2, f) various biochemical tests for identification of the bacterial isolates. Authors own images.

Figure 3. Details of a) and b) various biochemical tests of bacterial isolate 2, soil sample collected, c) and d) dye degrading capacity of the bacterial isolate 1), e) and f) dye degrading capacity of the bacterial isolate 2. Authors own images.

Figure 4. Dye optimization of a) bacterial isolate 1 at pH 4, b) bacterial isolate 1 at pH 6, c) bacterial isolate 1 at pH 8, d) bacterial isolate 1 at 37°C, e) bacterial isolate 1 at 33.5°C, f) bacterial isolate 1 at 40°C. Authors own images.

Table 1. Various biochemical tests of the isolated bacterial strains (isolate 1 and 2).

Tests	Bacterial strains	
	Isolate 1	*Isolate 2*
Carbohydrate test	Positive	Negative
• Glucose	Positive	Negative
• Lactose	Negative	Negative
• Sucrose	Positive	Negative
• Maltose		
Catalase test	Negative	Negative
Nitrate test	Negative	Negative
TSI test	Alkaline slant and acid butt	Alkaline slant and acid butt
Coagulase test	Negative	Negative
Oxidase test	Negative	Negative
Catalase test	Negative	Negative
Mannitol Motility	Non fermentative, non motile	Non fermentative, motile
Indole test	Negative	Negative
Methyl red test	Positive	Negative
Voges-Proskauer	Positive	Negative
Citrate	Positive	Positive
Urease	Negative	Negative

Table 2. Effect of pH and temperature at different concentrations using the isolated bacterial strains (isolate 1 and 2; n=3).

Conditions	pH			Temperature (°C)		
	4	6	8	37	33.5	40
Isolate 1						
Concentration 1	0.03 ± 0.01	0.02 ± 0.01	0.01 ± 0.01	0.16 ± 0.01	0.08 ± 0.01	0.04 ± 0.01
Concentration 2	0.02 ± 0.01	0.02 ± 0.01	0.04 ± 0.01	0.19 ± 0.01	0.06 ± 0.01	0.03 ± 0.01
Concentration 3	0.03 ± 0.01	0.01 ± 0.01	0.02 ± 0.01	0.05 ± 0.01	0.05 ± 0.01	0.04 ± 0.01
Isolate 2						
Concentration 1	0.00 ± 0.00	0.00 ± 0.00	0.01 ± 0.00	0.00 ± 0.00	0.08 ± 0.01	0.04 ± 0.01
Concentration 2	0.01 ± 0.00	0.00 ± 0.00	0.00 ± 0.00	0.00 ± 0.00	0.06 ± 0.01	0.03 ± 0.01
Concentration 3	0.00 ± 0.00	0.01 ± 0.00	0.00 ± 0.00	0.00 ± 0.00	0.05 ± 0.01	0.04 ± 0.01

Numbers represent means ± one standard error (SE) of the mean

5. Results and discussion

From the above morphological, microscopical and biochemical results the isolated organism was found to be *Bacillus sp* and *Pseudomonas sp* respectively.

5.1 Dye degradation studies

The isolated organism was inoculated into nutrient broth with 1% red and yellow colours and kept in rotary shaker and reading was noted at 490 nm.

The colour degradation was noted by OD and colour change. For red, colour change was from red to dark brown and for yellow, change was from yellow to green.

The *Bacillus sp* isolated in this study decolorized the dye with chromophoric group (azo bond in golden yellow). This may be due to the source (soil contaminated with untreated textile effluent) which contained dyes of various chromophoric groups. Members of the genus Bacillus have been reported to decolorize azo dye (Kalyani et al., 2008).

After taken the OD values, the percentage of degradation was calculated and from the result, it was found to be Bacillus sp has more degrading capacity than Pseudomonas sp. The most degrading colour was found to be yellow than red.

5.2 Optimization studies

From the above result the optimization studies was done with Bacillus sp with yellow colour varying with different concentration of dye with varying pH 4, 6, and 8 and temperatures 37°C, 40°C and room temperature and kept in incubation. OD was plotted at 490 nm. From the OD values obtained it was to found to be the PH 8 and temperature 37°C with the concentration of dye 4%.

6. Conclusions

Environmental pollution has been identified as a major problem in the modern world. One of the most important environmental pollution problems is the colour in water. Optimization studies were carried out and found to be *Bacillus sp* having more degrading ability than *Pseudomonas sp* with concentration of 4% having pH 4 and 37°C temperature. It can be concluded that Bacillus has more degrading ability and further studies can be done with *Bacillus sp* and can be used in a bioremediation process.

References

Acuner, E., & Dilek, F. B. (2004). Treatment of tectilon yellow 2G by *Chlorella vulgaris*. *Process Biochemistry, 39*(5), 623-631.

Ahmad, A. L., & Puasa, S. W. (2007). Reactive dyes decolourization from an aqueous solution by combined coagulation/micellar-enhanced ultrafiltration process. *Chemical Engineering Journal, 132*(1), 257-265.

Aksu, Z. (2005). Application of biosorption for the removal of organic pollutants: a review. *Process Biochemistry, 40*(3), 997-1026.

Aksu, Z. (2005). Application of biosorption for the removal of organic pollutants: a review. *Process Biochemistry, 40*(3), 997-1026.

Aksu, Z., & Dönmez, G. (2005). Combined effects of molasses sucrose and reactive dye on the growth and dye bioaccumulation properties of *Candida tropicalis*. *Process Biochemistry, 40*(7), 2443-2454.

Aksu, Z., & Dönmez, G. (2005). Combined effects of molasses sucrose and reactive dye on the growth and dye bioaccumulation properties of Candida tropicalis. *Process Biochemistry, 40*(7), 2443-2454.

Alaton, I. A., & Balcioglu, I. A. (2001). Photochemical and heterogeneous photocatalytic degradation of waste vinylsulphone dyes: a case study with hydrolyzed Reactive Black 5. *Journal of Photochemistry and Photobiology A: Chemistry, 141*(2), 247-254.

Anbia, M., & Hariri, S. A. (2010). Removal of methylene blue from aqueous solution using nanoporous SBA-3. *Desalination, 261*(1), 61-66.

Aravindhan, R., Rao, J. R., & Nair, B. U. (2007). Removal of basic yellow dye from aqueous solution by sorption on green alga Caulerpa scalpelliformis. *Journal of Hazardous Materials, 142*(1), 68-76.

Axelsson, J., Nilsson, U., Terrazas, E., Aliaga, T. A., & Welander, U. (2006). Decolorization of the textile dyes Reactive Red 2 and Reactive Blue 4 using Bjerkandera sp. Strain BOL 13 in a continuous rotating biological contactor reactor. *Enzyme and Microbial Technology, 39*(1), 32-37.

Bârsănescu, A., Buhăceanu, R., & Dulman, V. (2009). Removal of Basic Blue 3 by sorption onto a weak acid acrylic resin. *Journal of applied polymer science, 113*(1), 607-614.

Bârsănescu, A., Buhăceanu, R., & Dulman, V. (2009). Removal of Basic Blue 3 by sorption onto a weak acid acrylic resin. *Journal of applied polymer science, 113*(1), 607-614.

Bayramoglu, G., Altintas, B., & Arica, M. Y. (2009). Adsorption kinetics and thermodynamic parameters of cationic dyes from aqueous solutions by using a new strong cation-exchange resin *Chemical Engineering Journal, 152*(2), 339-346.

Brown, M. A., & De Vito, S. C. (1993). Predicting azo dye toxicity. *Critical reviews in environmental science and technology, 23*(3), 249-324.

Carliell, C. M., Barclay, S. J., Naidoo, N., Buckley, C. A., Mulholland, D. A., & Senior, E. (1995). Microbial decolourisation of a reactive azo dye under anaerobic conditions. WATER SA-PRETORIA-. 21, 61-61.

Carliell, C. M., Barclay, S. J., Naidoo, N., Buckley, C. A., Mulholland, D. A., & Senior, E. (1995). Microbial decolourisation of a reactive azo dye under anaerobic conditions. *WATER SA-PRETORIA-, 21*, 61-61.

Ceron-Rivera, M., Davila-Jimenez, M. M., & Elizalde-Gonzalez, M. P. (2004). Degradation of the textile dyes Basic yellow 28 and Reactive black 5 using diamond and metal alloys electrodes. *Chemosphere, 55*(1), 1-10.

Chang, M. C., Shu, H. Y., Yu, H. H., & Sung, Y. C. (2006). Reductive decolourization and total organic carbon reduction of the diazo dye CI Acid Black 24 by zero valent iron powder. Journal of Chemical Technology and Biotechnology, 81(7), 1259-1266.

Chatterjee, S., Lee, D. S., Lee, M. W., & Woo, S. H. (2010a). Enhanced molar sorption ratio for naphthalene through the impregnation of surfactant into chitosan hydrogel beads. *Bioresource technology, 101*(12), 4315-4321.

Chatterjee, S., Lee, M. W., & Woo, S. H. (2009). Influence of impregnation of chitosan beads with cetyl trimethyl ammonium bromide on their structure and adsorption of congo red from aqueous solutions. *Chemical Engineering Journal, 155*(1), 254-259.

Chatterjee, S., Lim, S. R., & Woo, S. H. (2010b). Removal of Reactive Black 5 by zero-valent iron modified with various surfactants. Chemical Engineering Journal, 160(1), 27-32.

Crini, G. (2006). Non-conventional low-cost adsorbents for dye removal: a review. *Bioresource technology, 97*(9), 1061-1085.

Dafale, N., Wate, S., Meshram, S., & Nandy, T. (2008). Kinetic study approach of remazol black-B use for the development of two-stage anoxic–oxic reactor for decolorization/biodegradation of azo dyes by activated bacterial consortium. *Journal of hazardous materials, 159*(2), 319-328.

Daneshvar, N., Ayazloo, M., Khataee, A. R., & Pourhassan, M. (2007). Biological decolorization of dye solution containing Malachite Green by microalgae Cosmarium sp. *Bioresource technology, 98*(6), 1176-1182.

Dávila-Jiménez, M. M., Elizalde-González, M. P., Gutiérrez-González, A., & Peláez-Cid, A. A. (2000). Electrochemical treatment of textile dyes and their analysis by high-performance liquid chromatography with diode array detection. *Journal of Chromatography A, 889*(1), 253-259.

Deng, D., Guo, J., Zeng, G., & Sun, G. (2008). Decolorization of anthraquinone, triphenylmethane and azo dyes by a new isolated Bacillus cereus strain DC11. *International Biodeterioration & Biodegradation, 62*(3), 263-269.

Dulman, V., Simion, C., Bârsănescu, A., Bunia, I., & Neagu, V. (2009). Adsorption of anionic textile dye Acid Green 9 from aqueous solution onto weak or strong base anion exchangers. *Journal of applied polymer science,113*(1), 615-627.

Duran, N., & Esposito, E. (2000). Potential applications of oxidative enzymes and phenoloxidase-like compounds in wastewater and soil treatment: a review. *Applied catalysis B: environmental, 28*(2), 83-99.

Easton, J. R. (1995). The dye maker's view. *Colour in dyehouse effluent, 11.*

El-Dein, A. M., Libra, J. A., & Wiesmann, U. (2003). Mechanism and kinetic model for the decolorization of the azo dye Reactive Black 5 by hydrogen peroxide and UV radiation. *Chemosphere, 52*(6), 1069-1077.

Erdal, S., & Taskin, M. (2010). Uptake of textile dye Reactive Black-5 by Penicillium chrysogenum MT-6 isolated from cement-contaminated soil.*African Journal of Microbiology Research, 4*(8), 618-625.

Fan, J., Guo, Y., Wang, J., & Fan, M. (2009). Rapid decolorization of azo dye methyl orange in aqueous solution by nanoscale zerovalent iron particles. *Journal of Hazardous Materials, 166*(2), 904-910.

Forgacs, E., Cserhati, T., & Oros, G. (2004). Removal of synthetic dyes from wastewaters: a review. *Environment international, 30*(7), 953-971.

Fu, Y., & Viraraghavan, T. (2001). Fungal decolorization of dye wastewaters: a review. *Bioresource technology, 79*(3), 251-262.

Galindo, C., & Kalt, A. (1999). UV–H 2 O 2 oxidation of monoazo dyes in aqueous media: a kinetic study. *Dyes and Pigments, 40*(1), 27-35.

Goheen, S.C., Durham, D.E., McCulloch, M., Heath, W.O. (1994) The degradation of organic dyes by Corona discharge.In: Eckenfelder, W.W., Bowers, A.R., Roth, J.A. (Eds.), Chemical Oxidation, v Technomic Publishing Co. Inc., Lancaster, Basel,2, 356–367.

Gong, R., Zhang, X., Liu, H., Sun, Y., & Liu, B. (2007). Uptake of cationic dyes from aqueous solution by biosorption onto granular kohlrabi peel.*Bioresource Technology, 98*(6), 1319-1323.

Gottlieb, A., Shaw, C., Smith, A., Wheatley, A., & Forsythe, S. (2003). The toxicity of textile reactive azo dyes after hydrolysis and decolourisation.*Journal of Biotechnology, 101*(1), 49-56.

Greluk, M., & Hubicki, Z. (2010). Kinetics, isotherm and thermodynamic studies of Reactive Black 5 removal by acid acrylic resins. *Chemical Engineering Journal, 162*(3), 919-926.

Gutierrez, M. C., Pepio, M., Crespi, M., & Mayor, N. (2001). Control factors in the electrochemical oxidation of reactive dyes. *Coloration technology,117*(6), 356-361.

Hao, O. J., Kim, H., & Chiang, P. C. (2000). Decolorization of wastewater.*Critical reviews in environmental science and technology, 30*(4), 449-505.

He, Z., Song, S., Zhou, H., Ying, H., & Chen, J. (2007). CI Reactive Black 5 decolorization by combined sonolysis and ozonation. *Ultrasonics Sonochemistry, 14*(3), 298-304.

Heinfling, A., Bergbauer, M., & Szewzyk, U. (1997). Biodegradation of azo and phthalocyanine dyes by Trametes versicolor and Bjerkandera adusta.*Applied Microbiology and Biotechnology, 48*(2), 261-266.

Ince, N. H. (1999). "Critical" effect of hydrogen peroxide in photochemical dye degradation. *Water Research, 33*(4), 1080-1084.

Iscen, C. F., Kiran, I., & Ilhan, S. (2007). Biosorption of Reactive Black 5 dye by Penicillium restrictum: The kinetic study. *Journal of hazardous materials, 143*(1), 335-340.

Kalyani, D. C., Patil, P. S., Jadhav, J. P., & Govindwar, S. P. (2008). Biodegradation of reactive textile dye Red BLI by an isolated bacterium Pseudomonas sp. SUK1. *Bioresource Technology, 99*(11), 4635-4641.

Kapdan, I. K., & Oztekin, R. (2006). The effect of hydraulic residence time and initial COD concentration on color and COD removal performance of the anaerobic–aerobic SBR system. *Journal of hazardous materials, 136*(3), 896-901.

Karataş, M., Dursun, Ş., & Argun, M. E. (2010). The decolorization of azo dye reactive black 5 in a sequential anaerobic-aerobic system. *Ekoloji,19*(74), 15-23.

Karcher, S., Kornmüller, A., & Jekel, M. (2001). Cucurbituril for water treatment. Part I:: Solubility of cucurbituril and sorption of reactive dyes.*Water research, 35*(14), 3309-3316.

Khehra, M. S., Saini, H. S., Sharma, D. K., Chadha, B. S., & Chimni, S. S. (2006). Biodegradation of azo dye CI Acid Red 88 by an anoxic–aerobic sequential bioreactor. *Dyes and Pigments, 70*(1), 1-7.

Krishnakumar, K. N., Rao, G. P., & Gopakumar, C. S. (2009). Rainfall trends in twentieth century over Kerala, India. *Atmospheric environment, 43*(11), 1940-1944.

Labanda, J., Sabaté, J., & Llorens, J. (2009). Modeling of the dynamic adsorption of an anionic dye through ion-exchange membrane adsorber.*Journal of Membrane Science, 340*(1), 234-240.

Lee, Y. H., & Pavlostathis, S. G. (2004). Decolorization and toxicity of reactive anthraquinone textile dyes under methanogenic conditions. *Water Research, 38*(7), 1838-1852.

Levine, W. G. (1991). Metabolism of azo dyes: implication for detoxication and activation. *Drug metabolism reviews, 23*(3-4), 253-309.

Lin, S. H., & Peng, C. F. (1994). Treatment of textile wastewater by electrochemical method. *Water research, 28*(2), 277-282.

Mahadwad, O. K., Parikh, P. A., Jasra, R. V., & Patil, C. (2012). Photocatalytic degradation of reactive black-5 dye using TiO2-impregnated activated carbon. *Environmental technology, 33*(3), 307-312.

Malik, P. (2004). Dye removal from wastewater using activated carbon developed from sawdust: adsorption equilibrium and kinetics. *Journal of Hazardous Materials, 113*(1), 81-88.

Manu, B., & Chaudhari, S. (2002). Anaerobic decolorisation of simulated textile wastewater containing azo dyes. *Bioresource technology, 82*(3), 225-231.

Martínez-Huitle, C. A., & Brillas, E. (2009). Decontamination of wastewaters containing synthetic organic dyes by electrochemical methods: a general review. *Applied Catalysis B: Environmental, 87*(3), 105-145.

Méndez-Paz, D., Omil, F., & Lema, J. M. (2005). Anaerobic treatment of azo dye Acid Orange 7 under fed-batch and continuous conditions. *Water research, 39*(5), 771-778.

Mishra, G., & Tripathy, M. (1993). A critical review of the treatments for decolourization of textile effluent. *Colourage, 40*, 35-35.

Morais, L. C., Freitas, O. M., Goncalves, E. P., Vasconcelos, L. T., & Beca, C. G. (1999). Reactive dyes removal from wastewaters by adsorption on eucalyptus bark: variables that define the process. *Water Research, 33*(4), 979-988.

Moussavi, G., & Mahmoudi, M. (2009). Degradation and biodegradability improvement of the reactive red 198 azo dye using catalytic ozonation with MgO nanocrystals. *Chemical Engineering Journal, 152*(1), 1-7.

Moussavi, G., & Mahmoudi, M. (2009). Degradation and biodegradability improvement of the reactive red 198 azo dye using catalytic ozonation with MgO nanocrystals. *Chemical Engineering Journal, 152*(1), 1-7.

O'mahony, T., Guibal, E., & Tobin, J. M. (2002). Reactive dye biosorption by Rhizopus arrhizus biomass. *Enzyme and Microbial Technology, 31*(4), 456-463.

Olukanni, O. D., Osuntoki, A. A., & Gbenle, G. O. (2006). Textile effluent biodegradation potentials of textile effluent-adapted and non-adapted bacteria. *African Journal of Biotechnology, 5*(20).

Osugi, M. E., Rajeshwar, K., Ferraz, E. R., de Oliveira, D. P., Araújo, Â. R., & Zanoni, M. V. B. (2009). Comparison of oxidation efficiency of disperse dyes by chemical and photoelectrocatalytic chlorination and removal of mutagenic activity. *Electrochimica Acta, 54*(7), 2086-2093.

Pearce, C. I., Lloyd, J. R., & Guthrie, J. T. (2003). The removal of colour from textile wastewater using whole bacterial cells: a review. *Dyes and pigments, 58*(3), 179-196.

Powell, W.W., Michelsen, D.L., Boardman, G.D., Dietrich, A.M., Woodby, R.M. (1994) Removal of color and TOC from segregated dye discharges using ozone and Fenton's reagent. In: Eckenfelder, W.W., Bowers, A.R., Roth, J.A. (Eds.), Chemical Oxidation, vol. 2. Technomic Publishing Co. Inc., Lancaster, Basel, 278–300.

Razo-Flores, E., Donlon, B., Field, J., & Lettinga, G. (1996). Biodegradability of N-substituted aromatics and alkylphenols under methanogenic conditions using granular sludge. *Water Science and Technology, 33*(3), 47-57.

Robinson, T., McMullan, G., Marchant, R., & Nigam, P. (2001). Remediation of dyes in textile effluent: a critical review on current treatment technologies with a proposed alternative. *Bioresource technology, 77*(3), 247-255.

Roriz, M. S., Osma, J. F., Teixeira, J. A., & Couto, S. R. (2009). Application of response surface methodological approach to optimise Reactive Black 5 decolouration by crude laccase from Trametes pubescens. *Journal of hazardous materials, 169*(1), 691-696.

Sanghi, R., Dixit, A., & Guha, S. (2006). Sequential batch culture studies for the decolorisation of reactive dye by Coriolus versicolor. *Bioresource technology, 97*(3), 396-400.

Sarioglu, M., Bali, U., & Bisgin, T. (2007). The removal of CI Basic Red 46 in a mixed methanogenic anaerobic culture. *Dyes and Pigments, 74*(1), 223-229.

Shu, H. Y., Huang, C. R., & Chang, M. C. (1994). Decolorization of mono-azo dyes in wastewater by advanced oxidation process: a case study of acid red 1 and acid yellow 23. *Chemosphere, 29*(12), 2597-2607.

Slokar, Y. M., & Le Marechal, A. M. (1998). Methods of decoloration of textile wastewaters. *Dyes and pigments, 37*(4), 335-356.

Sponza, D. T., & Işik, M. (2002). Decolorization and azo dye degradation by anaerobic/aerobic sequential process. *Enzyme and Microbial Technology,31*(1), 102-110.

Sponza, D. T., & Işik, M. (2002). Decolorization and azo dye degradation by anaerobic/aerobic sequential process. *Enzyme and Microbial Technology,31*(1), 102-110.

Stolz, A. (2001). Basic and applied aspects in the microbial degradation of azo dyes. *Applied microbiology and biotechnology, 56*(1), 69-80.

Supaka, N., Juntongjin, K., Damrongle¬d, S., Delia, M. L., & Strehaiano, P. (2004). Microbial decolorization of reactive azo dyes in a sequential anaerobic–aerobic system. *Chemical Engineering Journal, 99*(2), 169-176.

Talarposhti, A. M., Donnelly, T., & Anderson, G. K. (2001). Colour removal from a simulated dye wastewater using a two-phase anaerobic packed bed reactor. *Water Research, 35*(2), 425-432.

Tan, N. C., Borger, A., Slenders, P., Svitelskaya, A., Lettinga, G., & Field, J. A. (2000). Degradation of azo dye Mordant Yellow 10 in a sequential anaerobic and bioaugmented aerobic bioreactor. *Water Science and Technology, 42*(5-6), 337-344.

Tatarko, M., & Bumpus, J. A. (1998). Biodegradation of congo red by Phanerochaete chrysosporium. *Water research, 32*(5), 1713-1717.

Van der Zee, F. P., Lettinga, G., & Field, J. A. (2001). Azo dye decolourisation by anaerobic granular sludge. *Chemosphere, 44*(5), 1169-1176.

Vandevivere, P. C., Bianchi, R., & Verstraete, W. (1998). Treatment and reuse of wastewater from the textile wet-processing industry: Review of emerging technologies. *Journal of Chemical Technology and Biotechnology,72*(4), 289-302.

Vara Prasad, M. N., & de Oliveira Freitas, H. M. (2003). Metal hyperaccumulation in plants: biodiversity prospecting for phytoremediation technology. *Electronic journal of biotechnology, 6*(3), 285-321.

Vijayaraghavan, K., & Yun, Y. S. (2008). Biosorption of CI Reactive Black 5 from aqueous solution using acid-treated biomass of brown seaweed Laminaria sp. *Dyes and Pigments, 76*(3), 726-732.

Volesky, B., & Holan, Z. R. (1995). Biosorption of heavy metals. *Biotechnology progress, 11*(3), 235-250.

Wang, H., Zheng, X. W., Su, J. Q., Tian, Y., Xiong, X. J., & Zheng, T. L. (2009). Biological decolorization of the reactive dyes Reactive Black 5 by a novel isolated bacterial strain Enterobacter sp. EC3. *Journal of Hazardous Materials, 171*(1), 654-659.

Wawrzkiewicz, M., & Hubicki, Z. (2009). Equilibrium and kinetic studies on the adsorption of acidic dye by the gel anion exchanger. *Journal of hazardous materials, 172*(2), 868-874.

Wuhrmann, K., Mechsner, K. L., & Kappeler, T. H. (1980). Investigation on rate—Determining factors in the microbial reduction of azo dyes. *Applied microbiology and biotechnology, 9*(4), 325-338.

Xingzu, W. A. N. G., CHENG, X., Dezhi, S. U. N., & Qi, H. (2008). Biodecolorization and partial mineralization of Reactive Black 5 by a strain of Rhodopseudomonas palustris. *Journal of Environmental Sciences, 20*(10), 1218-1225.

Xiong, Y. A., Strunk, P. J., Xia, H., Zhu, X., & Karlsson, H. T. (2001). Treatment of dye wastewater containing acid orange II using a cell with three-phase three-dimensional electrode. *Water research, 35*(17), 4226-4230.

Yesilada, O., Asma, D., & Cing, S. (2003). Decolorization of textile dyes by fungal pellets. *Process Biochemistry, 38*(6), 933-938.

Yi, F., & Chen, S. (2008). Effect of activated carbon fiber anode structure and electrolysis conditions on electrochemical degradation of dye wastewater. *Journal of Hazardous Materials, 157*(1), 79-87.

Zaoyan, Y., Ke, S., Guangliang, S., Fan, Y., Jinshan, D., & Huanian, M. (1992). Anaerobic–aerobic treatment of a dye wastewater by combination of RBC with activated sludge. *Water Science and Technology, 26*(9-11), 2093-2096.

Zollinger, H. (1987) Color Chemistry: Synthesis, Properties and Applications of Organic Dyes and Pigments, VCH Publishers, New York, 92–100.

YOUR KNOWLEDGE HAS VALUE

- We will publish your bachelor's and
  master's thesis, essays and papers

- Your own eBook and book -
  sold worldwide in all relevant shops

- Earn money with each sale

Upload your text at www.GRIN.com
and publish for free